Margaret Virginia McCabe

Life Forces

Margaret Virginia McCabe

Life Forces

ISBN/EAN: 9783337779245

Printed in Europe, USA, Canada, Australia, Japan

Cover: Foto ©berggeist007 / pixelio.de

More available books at **www.hansebooks.com**

LIFE FORCES

Margaret Virginia McCabe

Author of "While I was Musing the Fire Burned"

Washington, D. C.:
Press of John F. Sheiry, 623 D Street
1899

NOTE.

No apology is made for the thoughts herein given. The Spirit leads and all must follow. A great truth is ready for acceptance; has been since the beginning. There are many ways to enter and each way seemeth best. "Trust thyself" is the best guide. As the Voice speaks to the inner consciousness, that is the road to take. Sickness abounds, and there is a mad rush to the pool as the waters are disturbed. Many stand back and complain, yet all are eager and waiting for relief, but they want a prop on which to lean and be pushed forward. The writer is not seeking to advertise her own gift of healing. She has avoided as much as possible all references to patients.

This little book is sent out to those eager, striving souls who are earnestly seeking aid on a rugged, toilsome journey. The light within its lines may be hidden to many, but to those for whom it is intended the radiance will shine forth and the cross will grow lighter.

It is meant also to relieve death of its horrors and teach the divine comprehension of the soul's progression. It is dedicated to none in name, but in thought and magnetic influence to many, and especially to suffering, saddened ones who need help.

The author is repaid by the silent thanks that telepathy brings back to her.

M. V. McC.

CONTENTS.

God-Consciousness 7

Material Forces 28

Spirit Forces 38

Building-Stones 57

Mortal Mind 64

Alpha and Omega . . 80

I.

GOD-CONSCIOUSNESS.

Words spoken vibrate through all eternity. Thoughts generated reverberate with greater force because the invisible powers are more potent in their results. Material forces cease with the death of the mortal body. Spiritual forces endure forever.

The first Word produced creation, and each succeeding vibration widened the circle by evolutionary processes. Back of this Word was a Cause, and that Cause was God—invisible because Spirit. We see the effect. We know and feel the cause. Human mind cannot comprehend God. Only through man's spiritual nature is the oneness felt.

Man is the effect of the Word. The soul man is God's individualization. External, mortal man lifted above the brute creation by his power of reasoning is simply the result of material forces.

Ideas photograph themselves on the brain.

Over one of our city colleges these words stand forth in glaring, glittering letters: "Education for Real Life." It sets one thinking. Real life! The tenement of clay, tossed about by every wind that blows, a creature of chance or man's pleasure. A life that ends when death calls! Perhaps we are mistaken and these are correct, but after years of experience in this real life I am ready to answer in the words of Phillips Brooks: "The ideal life, the life of full completeness haunts us all. We feel the thing we ought to be beating against the thing we are."

Once upon a time, in the far-off years ago, so long past that human mind cannot grasp its beginning, some mortal mind felt inspired to put together legends and weave them into a history of creation; a story of births and wars and bloodshed; a story of men of great physical strength, mighty and strong before the Lord. To these people many and devious happenings occurred, coupled with much wickedness and accompanied by visions of angels, and the world was filled with suggestions of ideas. Through all these centuries of strength, never was the people without one leading prophet guided by the Lord by means of visions. Finally, when the world grew more and more wicked the greatest prophet of all arose, and was called Jesus. Him they crucified, and ever since, the world of churches has worshipped the Man nailed to the cross. The shadow of that cross fell across the human race, and by suggestion, only the suffering, bleeding Jesus has been the world's Savior, and this thought has vibrated through 1900 years. A crucified Man has been worshipped, a man of sorrows and acquainted with grief, a humiliated, despised Jesus! But the risen Christ, victorious over evil, with His Father's seal upon his forehead shining with an inextinguishable glory, holding forth peace and comfort to the sorrowful race, has been ignored. The suggestion of the shadow of the cross has shut out the brightness of the all-comprehensible light—which is Love. And yet we call ourselves an intelligent, far-seeing race! We have dwelt so long on the shadow side of life; we have daily climbed Sinai's and knew it not—because we enjoy sympathy. We want to be sad, we want to be sickly, we want to be burdened and depressed, so we do not suffer too much.

Tear aside the close-shut gates and step into the broad, illuminated sunshine of God's love and worship the risen Christ. Ascend into the holy of holies, the inner sanctuary of your own God nature, and be at one with God himself.

There are so many discrepancies, so many contradictions in this book that men have written and called the Bible. Not a page that does not contradict itself. Even the thoughts of this man Jesus are changeable. The world has bowed down and done reverence to the mortal part of the prophets. The inspiration of the Spirit, that might have set upon them as cloven tongues of fire, has been ignored, and when one rises up and asserts this ego now, he is scoffed at as mad. The Lord said, "My Spirit shall not always strive with man for that he is also flesh."

The Bible is the product of mortal mind, historically incorrect, full of allegories and visions, but also full of inspired words. It pictures a God of anger and vengeance, smiting and killing and working His own selfish ends. How erroneous when God is only Love!

This terrible God is a God of mortal conception, builded by fear, and exists only through perverted, ignorant, evil suggestion. Ingersoll rightly says, "The devil is the keystone of the arch of Christianity," because the devil is an interpretation of mortal mind as a compensation for the avenging God man also created.

The Bible has come down through many ages because it had among its chaff inspirations of the living fire.

For man to abase himself, crucify his inclinations, and follow his own misconceived idea of the lowly Jesus, is simply ignorant belief. Jesus never humbled himself. He always exalted himself, and taught as one having authority. From His childhood's days, when He argued with the wise men in the temple to the hour He rebuked his disciples for murmuring over the loss of the costly box of ointment, you cannot find any degree of lowliness of mind or position. He asserted the God at all seasons and in all places. When they came to Him and said, "Your mother and brethren await without," was this Man touched with pity when, looking at His disciples, He replied, "These are my mother and brethren." How did those awaiting feel at thus being ignored?

Bring the application to our own homes and comprehend the lowly Jesus. The history of this man was written hundreds of years after its occurrence. Bring this historical or allegorical Jesus into our own time.

Would we follow blindly, or call our reasoning powers into question? Which? But take the risen Christ, the bright illumination of the heretofore misunderstood God—the word spoken into a psychic atmosphere —the perfection of the ideal life—when mortal mind with its selfish, ignorant beliefs has been crucified and the God, the Creator, the Spirit of Truth reigns, controls both the spiritual and material forces, then the perfection of living—the Comforter has come.

God who commanded light to shine out of darkness hath shined in our hearts to give the light of the knowledge of the glory of God in the face of Jesus Christ.

Matthew begins his gospel with these words:

"The book of the generation of Jesus Christ, the son of David, the son of Abraham," through forty-two generations, when Jacob begat *Joseph*, the husband of Mary, of whom was born Jesus, who is called Christ. Now, the birth of Jesus Christ was on this wise; when as His mother Mary was espoused to Joseph, before they came together, she was found with child of the Holy Ghost. If Joseph the husband came of the house of David, but knew not Mary till she had brought forth her first-born son, who was Mary, and how could the child born of the Holy Ghost belong to the house of David? The plea of kinship will not answer our reasoning mind. Why not give Mary's geneology instead of Joseph's? In those far-off days women were the silent ones, and Jewish history records only the paternal line, yet Jesus was no child of Joseph, in a material line, but tracing back the fingerprints in the sand the one universal truth comes to the front, even the evolutionary process of psychic development—the prophets, patriarchs, and kings—women dethroned if it pleases Jewish history— we reach the great cause and the one source of all life— God. Jesus—historical or allegorical—was God's individualization through Mary as an instrument, whether she was known by Joseph or not. Leaving out the question of the Immaculate Conception, because all legends, all mythological ideas are born of the miraculous, and the question does not pertain to the soul's consciousness, but to reasoning mind—what did Matthew mean? Why was it necessary to show the lineage and raise a question at the climax over such a material, unimportant point? Would we believe and accept such wonders to-day and follow so lowly a Jesus? No.

John is more spiritual in his conception, for he deals not with this part of mortal belief, but says:

"In the beginning was the Word, and the Word was with God, and the Word was God. The same was in the beginning with God. All things were made by Him; and without Him was not anything made that was made. In Him was life; and the life was the light of men. And the light shineth in darkness; and the darkness comprehendeth it not."

Which seems the inspired writer?

Matthew deals with the material, mortal man, and is filled with errors.

John catches the conception of God and writes in touch with the infinite.

Matthew shows us Jesus.

John teaches us God.

"He was in the world, and the world was made by Him, and the world knew Him not."

The prophets knew Him, and Jesus knew Him. The apostles knew Him through Jesus as a medium, and their works were limited by their faith.

Moses recognized Him in the burning bush. Elijah heard Him in the still, small voice that spake not in the tempest or the earthquake, but came in the silence.

What is now has always been. God is the Alpha and Omega—God is the Word, and when He spoke creation began.

The vibrations have widened century after century. The same psychic atmosphere exists above the clouds of earth that existed in the beginning. God cannot change. Everything else changes by suggestion, that has grown into custom.

"The mists came up and covered the earth" and shut out God.

There was perfect harmony in the beginning until evolutionary process, with its creative energy, brought to perfection its physical development, Man—with his dual mind, the objective or reasoning mind shutting out the subjective or intuitive.

Fisk says: "The primal origin of consciousness is hidden in the depths of bygone eternity." He also says: "Universal struggle for existence having succeeded in bringing forth that consummate product of creative energy—the human soul—has done its work and will presently cease. In the lower region of organic life it must go on, but as a determining factor in the highest works of evolution it will disappear."

Why?

Because evolution must proceed within man himself.

The sons of man must evolve into the sons of God. When the human soul has awakened to its environments it will vibrate into its own vast eternity.

There must be friction, because we do not always comprehend the struggle. Jesus was cognizant of this friction and He prayed, "Father, if it be possible let this cup pass from me." But it was not possible.

Soul cannot come into its own but by crucifixion of its mortal consciousness, but when the dark hour is past and the resurrection is assured, let us not live in the shadow of the cross, but beyond in the illumination. Carry the transfigured, risen Christ to suffering fellowmen.

We are so tired of the shadows—so weary of the endless wounds of agonizing pain that the ages have laid upon us. We cry out with strong heart, yearning for the sunlight of Love to fall across our weary lives. We have suffered so long, and the pain seems never to have run its length. Give us peace, love, and sunshine—and it is ours for the asking.

Fisk, moreover, says: "Through misery that has seemed endless, men have thought on that gracious life and its sublime ideal and have taken comfort in the sweetly solemn message of peace on earth, good-will to men."

This is God's message through his inspired medium— and it covers all the Gospels. If we are not at peace with our neighbors life with us is out of shape. The circumference has sprung a tangent. One God; one creative force; one spirit pervades everything.

He is the same God that Adam recognized, that spoke through Moses, Elias, John, and Jesus. Each one interpreted Him differently, according to the signs of the times.

Moses' dispensation was amongst an entirely different set of people than those Jesus ministered unto. Each century has created its own ideal of God, but at no time has any man seen God. "While my glory passeth by I will cover thee with my hand—but my face shall not be seen."

We only see

"What feeble lenses and weak sight may scan,
And thus a double lessening, double veiling
Of the unimagined glory of a thought of Him
Who dwells between the cherubim!

> " The vision
> Which angels might receive straightway
> Unshorn of any ray,
> And hold in full possession,
> Must enter by the portal
> Of faculties—sin-paralyzed and mortal. "

And God always appeared to his children in visions. He took not away the pillar of cloud by day—nor the pillar of fire by night. He led them by devious turnings. Often they had to retrace their steps, and the labyrinth seemed endless, but mortal mind must battle for supremacy, and by repeated falls and beseeching cries rise again to the illuminated path.

The temptation of Jesus was this same struggle.

"At length there came a wonderful movement—silent and unnoticed as are the beginnings of all great revolutions," and now the world of the present day is starting into understanding of the grand conception of the Spirit that breathed o'er Eden.

> He walks in the earth and the heavens,
> The Lord in His raiment bright.
> His robe is crimson at evening,
> It is gold in the morning light,
> And it trails on the dusky mountains
> With a silver fringe at night.
>
> High over the people thronging
> Is the light of His pure face.
> Can the uttermost need and longing
> Come fronting that awful place?
> But to touch the beautiful garment
> Is a comfort and a grace.

He turns, and I am not hidden,
  And He smiles and blesses low.
Did the gift come all unbidden ?
  Oh, to think He would not know,
Though even the hem of His garment
  It was faith that touched Him so.

By Moses came the law, but by Jesus came faith.   In the early records there is no talk of sickness—healing arts were not necessary.   "Whatever theory of creation we may hold, we must believe that in the mind of the Creator was a pure and perfect ideal."   Moses was the lawgiver and the prophet that spake face to face with God.   We find murmurings and discontent all through the wanderings of the children of Israel, but we find death coming only through age.   No talking of sickness or disease, only disease of harmony; a decrease according to the laws of progression, for nothing human remains. If we do not increase in the strength of God, we decrease along the lines of mortal supremacy.

Jesus' ministry was one of healing.   Sickness filled the land.   The people kept not the covenant, but lost the balance between God and man.   Many strange doctrines had sprung up, and by evil suggestions physical suffering became paramount—the land was defiled— then came Jesus healing and teaching.   Not to build churches or establish creeds, but to lift a cursed nation above the surroundings of its own diseased imagination. And allegorically He taught them that by finding the kingdom of heaven all else should be added.   But the kingdom of heaven was not external, not a far-off, visionary place, but within, and the way to open the gates was to leave all, possessions, seats of custom, kindred, and by the crucifixion of mortal self, vain, ignorant, autocratic, deluded self-consciousness; and after the three dark days, when the earthly ties should quake through the friction of self, the veil of the fleshly temple be rent—then should we rise illuminated in the God-consciousness—and all mortal enemies be vanquished, disease should be no more, because God is perfect and has come into his kingdom, and we are no more subject to the law because we are born again of the fire and water.

Stars no longer fight in their course against Siscera, because the spirit hath made thee free. Free to enjoy your birthright here on earth. Free to receive your blessing even though your mess of pottage hath slipped from your grasp. But, having received the blessing of the anointing, you must not shut up the gift within self, but let your illuminated soul countenance shine before all men. Nay, you cannot hide it, for it will go before and illumine your path, and all power will be given you, power over all forces, visible and invisible.

By concentration the desires of your mind will go forth like the dove and return with the olive branch.

What is concentration? The shutting off of all material, external sense and dwelling in the holy of holies, the inner sanctuary—the opening of the sixth sense, or the inner man—the God plane--and where God and angels dwell the place is rarefied.

Moses approached this plane by laying off the material sense, expressed as taking off his shoes. Elijah met God in the stillness of the cave—shut in from mortal thought.

Is this power easily accomplished? Very.

Not suddenly, but by power and might, and steady purpose, unflinching desire, and patience.

You may start for the land of Canaan, but the flesh-pots of Egypt will often call you back, and weary of the noonday heat you may sit down by cooling streams and hang your harps on the willow trees, but the morning light will find you pressing onward, for having started there is no turning back, and there is much of your past life to unravel. Only by perseverance can you ascend into the mount of transfiguration. Nor can you long abide there for your work lies in the vale below.

Clairvoyance and clairaudience are attributes of this inner self. Were I to put in words the strange visions that come to me through this unfoldment, the material world might deem me insane. At first, the wonders were so great I consulted a celebrated M. D., and asked if I were fitting myself for an enclosure of stone walls. I even went to a phrenologist, and satisfied my mortal pride.

I was like Thomas; I wanted to put my fingers in the wound-prints—and I did.

You ask, did I suffer during this crucifixion? The answer lies too deep for words. Because I was so blindly egotistical on an intellectual plane, I thought I did not need God, and I shut him out of my calculations.

So the war went on, and I had to learn of sorrows and griefs. This great reasoning, mortal mind must know the why and wherefore before I learned to look within. How did the end come? There was no end, simply the beginning. I gave up and said, "lead me; teach me." Two graves had to be opened before the truth came to me. I had to go to the very border land before my crucifixion was accomplished—but while I was musing the fire burned—and now I faint not in the days of adversity.

"For what is Mine shall know My face."

What is Telepathy? Soul communing. Is it a common attribute? To some, yes. To others, faintly perceptible, but not understood. You can't reason over it. You must just simply accept blindly and await development. The answer may come by clairaudience or by clairvoyance. It depends upon your psychic development.

Listen to Cornelius Agrippa on this subject: "There is an art known only to a few, by which the purified and faithful soul of man may be instructed and illuminated so as to be raised at once from the darkness of ignorance to the light of wisdom and knowledge. If the soul is perfectly purified and sanctified she becomes free in her movements. She sees and recognizes the divine light, and she instructs herself while she seems to be instructed by another. In that state she requires no other admonition except her own thought, which is the head and guide of the soul. She is then no more subject to terrestrial conditions of time but lives in the eternal, and for her to desire a thing is to possess it already. Man's power to think increases in proportion as this etheral and celestial power of light penetrates his mind, and strengthening his mental faculties, it may enable him to see and perceive that which he interiorly thinks, just as if it were objective and external.

"Spirit, being unity and independent of our ideas of space, and all men having, therefore, essentially the same spirit, the souls of men existing at places widely distant from each other may thus enter into communication and converse with each other exactly in the same manner as if they had met in their physical bodies. In this state man may perform a great many things in an exceedingly short period of time, so that it may seem to us as if he had required no time at all to perform it. Such a man is able to comprehend and understand everything by the light of the universal power of guiding intelligence with which he is spontaneously united."

Many theories are advanced for the sake of this great science, but no one has laid down a law by which any one can work. Hudson has given us a clearer understanding of this great subjective mind—as separate and distinct from the objective mind, but connected by auto-suggestion. We have all grasped Hudson's teaching as a drowning man catching at straws, for he gives us a working hypothesis, but he doesn't tell any one how to obtain the key, because the development lies in self. You must work out your own salvation. First find your own kingdom—and so I abrogate to myself no power when I grasp the great God-consciousness. That is all-pervading. Crucifying mortal mind, I let the illuminated soul follow its own direction. I myself am nothing, for God is all. The spirit forces that dominate the soul plane must war against the material forces that rule the physical plane until the veil of the temple be rent.

Spirit being everywhere and mortal man only a temple for this spirit's individualization, when by concentration I am able to rise above the realization of the physical—the spirit being only a part of the vast and one great God—I am able by auto-suggestion to call to my inner sense perception what I desire, or else to send in any direction my loosened or freed spirit.

Results differ according to material bodies. For instance: I wish to reach a distant friend; I succeed; he is conscious of my presence, of my thoughts, of my acts—perhaps he thinks he is only dreaming, consciously or unconsciously—I know better because he writes of these dreams and thoughts and they are just what I willed. I am conscious at the time only by impressions, the material effect; but I received those letters and read them by my spirit's understanding even before the postman leaves them at my door. This person is a materialist on whom I experiment without his knowledge.

But let me draw toward me one versed in psychic knowledge and the result is different—then I see; I can clasp hands and feel the touch of flesh, living flesh; and I can converse and receive replies.

No, the other psychist gets only impressions, because his soul is beyond the body and has come to me—and unless he suddenly awakes and recalls the power, the knowledge does not rise above the threshold of conscience.

For mere curiosity, I began, by request, telepathic communication with another party with whom I was not *en rapport*. He was a reasoner and I don't know just what he did expect, but the matter was a failure. Why? Because I had no interest in satisfying mere curiosity and I was not attracted to or by that sphere. Our circles did not combine, but this I assert as a fact, demonstrable to my own satisfaction, that when I am interested and pleased to experiment I can and do hold communion with kindred souls by telepathy.

Another fact worthy of study is this—that the best results are obtainable when experiments are made between the sexes—male and female are complementary and complete a sphere.

By telepathy, all mental and medical cures are affected —drugs do no good unless accompanied by faith—they may relieve the effects of disease but they do not remove the cause, and while they may build up one portion they tear down another—but mental healing goes direct to the cause, eradicates the error, and allows life to course freely through its channels. Through the power of telepathy, suffering humanity can be lifted above its sorrowful, saddened plane of existence into the free, pure air of the sunshine of health.

A patient in western Missouri whom I have never seen, says she can often feel my presence near. Once I wrote her, "Silently hold my letter near you ; the magnetism will permeate your body and make you feel rested and comforted all day." She replied that she did as directed, and in the evening changing her woolen wrapper the electricity emanated in all directions with a hissing, burning noise, and also her felt slippers were so charged—a very common occurrence where woolen goods play the part of conductor—*but she had never noticed it before.*

Was this suggestion? Perhaps. But what generated the unusual electric disturbance?

Was it vibrations? Yes.

Once this quotation expressed my views.

"Out in the silent dark I grope alone
    With human fingers tangling up the threads
Of God's eternal truth. Sullen, I moan,
    The light is for the dead—blessed dead."

Now I know differently. I had an aged and enfeebled mother under my care for a few short months. She lived entirely on my mentality. I thought and acted for her, but I knew she must soon pass over, because she wanted to go and be with the father, who had preceeded her by only a few months.

My dear friends, you must feel the vibrations from my pen as I write these words in all tenderness and love. Not in science, for I gave blindly. My crucifixion had not been finished.

For some days before her spirit departed I felt an awe, and I nightly dreamed dreams, and felt the perfume of flowers we use around the dead.

At nine o'clock one night the summons came, and as I picked her up and laid her on the bed (she had been down stairs a few short hours before) I knew I had to face my darkest hour—and self was lost in the God-consciousness.

By artificial means employed by two physicians, life was kept in the poor, tired body for thirteen hours, hoping loved ones from a distance might reach her, but failed in the end.

For those thirteen hours I sat by her side giving her all my vitality and magnetic force, with this one suggestion ringing itself through my brain, "Go, Soul, since go thou must, but go without a struggle." She was conscious all the time. The medicine they poured into her never affected the physical system. It had run its

race and the outgrown shell was to be left, but the soul
was shining clear and bright. " No pain," she said,
" only tired."

I try so often to imagine that feeling—the tired body
dropping away from the soul. No pain—only tired !

As I saw the end drawing nearer and nearer I put my
open palm on her forehead and with the other hand
clasped the tired fingers that would so soon be folded over
the quiet heart—life's labor done—and let all the vitality
flowing through my body be given her, and as the sun
poured in the window, and the birds took up their ma-
tins, quietly, peacefully, sweetly, the breath came slower
and slower, and without a struggle, without a murmur,
the mortal eyes closed and the inner sight was opened.
To the opening of the door forever have I been, and if
I stand without and knock, oft times a ray comes to
mortal ken, and I know and feel I have held communion with the dear ones on the other side.

Whether my hands held the magnetic power then, or
whether they gathered in those hours the blessing of the
spirit—I know not—I wish I did—I only know this
power has come to me.

To come face to face, for the first time, with death
leaves an impression of awe and stateliness that time
nor place can ever obliterate, and to live over those
scenes in thought is to return in spirit.

" In the light of the King's countenance is life."

Some one lately asked me what I believed regarding

the future life. I have no belief—I do not care. It does not affect me. I only know I do not believe in the material heavens the Pulpits declare. If we do have to walk golden streets, or sit down and play forever on golden harps, it will be very monotonous, and very glaring.

I have begun my progression here, and my work lies here until the soul grows tired of this physical frame, when it will lay it aside and continue its course—but the end is not yet, and what there is to learn I stand fearlessly and eagerly "amid the eternal ways" ready to learn, and when the door opens for me I shall just as fearlessly and just as eagerly take up the work beyond.

"Or ever the silver cord be loosened, or the golden bowl be broken, or the pitcher be broken at the fountain, or the wheel be broken at the cistern. Then shall the dust return to the earth as it was, and the Spirit shall return to God who gave it."

II.

MATERIAL FORCES.

The mists that have enveloped the earth have shut out the simple glories of all that is and all that ever will be. We stretch with yearning vision for the invisible, and our cries return to us with a faint echo of the minor chords of despondency. We want something— we know not what. We are tired of the creeds of men. We have found them wanting. Life drags itself out with pitiful, sad dejection. We are tired. We have striven for the unattainable and have found nothing. We awaken with a strange longing in our hearts, and we ponder over the mysteries of life, and are not satisfied. All is so dreary, so unsatisfactory. We read books and they soothe for the time, but they teach us the unreal. We try friendship, but we find it galling because envious and one-sided. One gives, the other takes, and the friend turns our enemy at last.

The broken pitchers of the Midianites, a few lights gone out, and the cry of fear goes up, and we turn and rend each other !

We try love, but it makes us more restless than ever. We are getting nearer the truth. We are finding ourselves in another. We are nearer completing the sphere.

But after a time love, too, proves fallacious. Love is so exacting. It wants all, and we have not all to give. We feel a loss, a hesitancy, a spot untouched by mortal ken. Where is it, and what is it? We feel the beating, pulsing life against the bars of mortality, but our wills are invincible. They must know the why and wherefore. Oh, the inexpressible bliss of closing our eyes to the blind, ruling, scheming science of life's material round and drifting on the tide of ideal reality! Not to struggle! Not to know! Not to care! Just to dream dreams and see visions, and be at peace with our lives. The impractical side of life hems us in with walls of impenetrable thickness. If we make one little chink, after years of labor, and let the starlight of dreams shine on us we are censured, and unless we stand, strong in our own strength, and care not for the blasts of the mighty or the mud of the lowly, we must go down to greater depths because we have gained one little vision of the truth, and if we turn back we are damned.

There is no turning back. Begin climbing, and you must continue. Do not expect to reach the top. The time is not yet come, but the more surely we build, the higher we get, the easier will it be for another to climb above our footprints. Nothing comes to us but for the good of all. Nothing happens to one but the entire race is benefited. My disappointments are so meant. O, yes, they are hard, and I feel the fret and worry they produce, but I find myself always questioning what the result will be. Pain always brings profit if we accept it in a kindly light.

I know days of lingering suffering—when between the paroxysms of pain the sweetest thoughts have come to me unbidden, and each time the pain grew less.

The lesson was kindly given, but we are so slow to learn. We want so much, but we want it all right now, and our weak strength is not sufficient to receive so great a gift. We must grow into the fulness.

I am feeling the thrilling response of a great truth—but I have been years learning even a little—but the soul has awakened and now more surely as I go I find doors opening and the way clearing.

No, it is not one green stretch of level land flowing with milk and honey. I cross mountains and I wade through deep, tempestuous waters, and often and often I faint by the wayside, and the grasshopper seems a burden, and the desire to push forward seems cold and lifeless, and I question everything, but my own forces pick me up and push me forward and greater strength comes with every failure. Only I count nothing a failure since all means a way to the end.

When we love we love blindly, and then some sad day we realize our mistake. One has mounted higher, and although we long and strive with all our soul to draw the loved one with us it is dead sea fruit—and in our bitterest hour the soul is alone. We are meant to rely on self—the ego—and not lean against another.

The saddest word in the English language, and yet the strongest, is that word alone. With none to understand, none to sympathize, none to feel the longings, yearnings! Thrown on our own resources we stagger and grasp for some prop to lean upon—but there is none.

Then comes the blankest hour of life—to sink or swim!

We are so weak, so bruised and battered by winds of adversity—but we conquer by our own little efforts and strength comes, and by and by the victory, for we have conquered forever, and alone we stand.

Alone with God.

Nothing but good gravitates toward us—but we had to pass through blackest torment—for our power.

And it is ours, now and forever more. And the loneliness and sadness is gone. For all things are ours by the power of Love.

Vain imaginings of our own unlimited desire. What is wrong? Have we missed the right interpretation of Truth, or are we on the wrong road after all? Is it better to sit down in the shade of the valley—the pale, purple light of the rainbow tints—and let mortal love cloud our eyes to the glorious, radiant beams of the mountain height of sunlight?

Poor soul! Not to know that having started on this upward climb often will you look back with sad and weary eyes to the monotony of the shadow, each day growing weaker and wearier of the aimless existence. You may look back. You may even take one step backward, but there is no returning, for as you look up once more to the eternal sunlight you long more and more to reach the summit where the prismatic tints are blended into one great white light of eternal radiance.

Soul must run the entire gamut.

All was darkness, desolation, and despair in the valley below before the evolution began; then the purple light of Love opened the Soul's eyes to the first dawn of day, then step by step, mistake after mistake, error upon error—tears more bitter than thought can imagine—it climbed the stairs of silence and came into its own and beheld the glorious Light, the Light that never was on sea or land.

People come to me and say, we cannot find the way, we have read until we are tired, we have tested every experience and we find no law to guide us, and we are tired of the struggle.

Then I say, oh, my friends, we may put you on the way, we may point it out to you, but you must build for yourself. I may take you over my tracks but they will not satisfy you because it is not your own way. Spirit is the same with all, but the mortal bodies are builded differently from each other by preconceived thought, and we may all reach the same goal, but our ways will deviate on the journey.

I sit and watch the hum of life go by. Here a wagon

with horses slowly dragging their burdens—now a hand-
somely equipped carriage—now a man riding a wheel
—the cars ladened with their human freight, some of it
misshapened and dwarfed, some of it full of vitality and
mentality, and some ladened down with the sadness of
mortal failure.

I sit and muse on the little lives of men and how we
mar them by our envious, jealous, unkind thoughts.
What are they living for? Future reward; and what
will it be?

We build our own ideas of heaven, and some picture
a hell !

I want no greater, more horrible idea of hell than I
have experienced right here on earth—and the heaven
of our dreams will be disappointing.

What does the future mean to me or what do I care
for the heaven I have had thrown at me while the present is one howling bitterness of woe !

I want to live now and enjoy this existence without
questioning the future for which I do not care, conscious
only that the greater progression I make now, the better
Eternity will be.

Some one said to me the other day, "you dwell so
much on God." Why not? God is all that is; and
what God is, I know not. The Bible did not teach him
to me and for years and years I feared the thought of
this avenging, angry being who would create a hell and
a devil and send his own creations into this everlasting

punishment. So I left him alone—put him out of my calculations, and, as I thought, builded my own life—alone. Physical suffering opened the spiritual vision and this God began to show himself to me. In the people that pass my windows, in the dumb animals, in the birds, the flowers, the grass, in the mighty rocks strong in their quietness, the ever restless sea, the faint streaks of dawn, the splendor of the noonday sun, and the trailing clouds of glory of the after glow!

I find Him in the sections of an orange, of every fruit that grows, and every flower, arranged with mathematical precision.

What is this God? I know not, only I cannot fear what is myself. I see Him all around and within. I link myself with His intentions and I feel the increasing vibrations of His love because there is nothing else to know. This creative force has individualized itself through my mortal body, and when the mortal race is run my eyes shall be opened clearly to what has been all the time.

Death makes no change.

I am weary of the fallacies of mortal mind—its narrowness, bigotry, idolatry, and ignorance; where nothing lasts, nothing remains right, and nothing stays wrong. Changes always! and we grow no nearer true comprehension. I am weary of it all!

For many years I was weary of pain, of sickness. Drugs relieved for a time perhaps, but the pain returned.

Then came one great agony, and two ways were pointed
to me; one of invalidism, or I was told, like Mrs. Dom-
bey, "make an effort," and this from a scientific medical
man!

So I turned my back to the wall and got up, intend-
ing to find a surer means than this uncertainty or end it
all! And so it has ever been. The law shuts us in and
idle gossip hurts, and we live for the criticism of the
world.

Not to know that the only law is the natural law!

Not to know that the discords of inharmonious life
are simply the surroundings of our own fretful, worried
nature!

Nothing harms me but what I, myself, do.

Nothing can touch me but as I invite it.

What I give out I get in return.

Simply a very plain way of living the Golden Rule.

People are continually knocking their heads against
brick walls. The knock hurts themselves only; the
wall stands unmoved. The soul is unconquerable. It
stands alone, and the harder the blows the stronger it
towers above the bulwarks of weak, erring, blinded human intellect.

> "Out of the night that covers me,
>   Black as the pit, from pole to pole,
> I thank whatever gods there be
>   For my unconquerable soul.

"In the fell clutch of circumstance
   I have not winced nor cried aloud.
Under the bludgeonings of chance
   My head is bloody, but unbowed.

"Beyond the place of wrath and tears
   Looms but the horror of the shade,
And yet the menace of the years
Finds and shall find me unafraid.

"It matters not how straight the gate,
   How charged with punishment the scroll;
I am the master of my fate,
   I am the captain of my soul."

This little poem helped me over a rough hour once, and I give it now, with my own thought vibrations, knowing it will help others. Some poems, some thoughts are just as much inspired, to my thinking, as ever the words of the Bible. Even more so, for I comprehend better the feelings of humanity now, than I can go back and pick up the thoughts and inspirations of those writers of the Bible, because their material surroundings, social laws, and customs, surrounded by the mysteries of the East, were not as our surroundings.

But I seem to multiply words and yet repeat always one Truth—the all-conquering Truth of the ages—that not around us is the means of faith, not in idly prepared words of salaried men of God—though at times we catch the inner vibrations of a truth they dare not utter because their Creed forbids!—but within is the fountain of Light, and only by the inner vision of sin-paralyzed mortal vision, can we find what we seek so perseveringly—but not patiently—in the external world—and find not.

By introspection all gates are opened, all ways are cleared, all discords abolished, all pain healed, and divine, complete, diatonic harmony, resounding through all the ages is found, lying dormant, only awaiting recognition.

The Kindom of Heaven is within.

It is the Kindom of Love—and Light.

Finding this, all else is found.

III.

SPIRIT FORCES.

In dealing with this subject I am perfectly aware I am bringing down on my head much criticism.

I am not seeking to establish any theory, but to deal with such facts as have come to my personal knowledge.

It is a subject of widespread interest and yet of much doubt. There are hours when I lose faith and call it all imaginations of a supersensitive brain, but the reality of the truth forces itself upon my doubts and will not be put aside for a more convenient season.

I firmly believe that the cloud of witnesses in the unseen world are close around us, leading us for our good, and our darkest hours are the seasons when mortal mind wars against this control and the inharmonious atmosphere admits disease. The sooner we accept this fact the quicker will be our freedom.

Horoscopes may be cast and books on astrology be lavishly printed, but Eleanor Kirk has said one true thing if nothing else be said, and in words to this effect: "'The signs of the zodiac may rule our mortal destinies, and we all find similarity in the rules laid down. For instance, Pisces foreshadowed my early years quite closely until I found many phenomena outside the precepts, but could not comprehend, until she nobly acknowledged the discrepancy and says, one born into the Spirit towers above all material surroundings and makes and controls its own destiny, and then the signs fail." So I lay aside my horoscope. A phrenologist examined my head; that, too, I outgrew. Palmists and gipsies read my life line; they, too, proved fallacious. Planchettes and table rappings and furniture knocks never interested me. It seems so absurd that the spirits of departed ones need condescend to material, narrow surroundings to communicate with loved ones when the one great mind is all the universe!

Not to know that one's own thought is sufficient to produce any phenomena! Absurd!

Wonderful and startling communications have been given—and learned men look and are astonished, and even students of Hudson's Law of Psychic Phenomena put aside these questions and confess themselves baffled.

The subjective mind knows all; past, present, and future; whether we have sufficient development to draw the knowledge above the threshold of consciousness or not, yet it works out its own way, even if that way be by raps or knocks—in other words, it is Telepathy.

This law is my foundation for my own experiments, and I give them through no sense of egotism or desire to assert a law for others; only this, by steps that each individual makes, the race may better climb.

Perhaps, being a psychic with a strangely developed

nervous system, I may get stranger results than others.

I have not completed my investigations and never expect to, and I am open to convictions.

Between Spiritualism and Spiritism there is a wide difference, in my estimation. I will first take up Spiritualism—a very interesting and wonderful power—and, if the conditions be favorable, easily acquired.

Now, this is my belief—which I again repeat—and it is from Hudson.

My subjective mind being my storehouse of memory—and the seat of all emotions—knows everything; past, present, and future; things I may have known and forgotten, or things my objective mind may never have been conscious of. Time and space being not considered, this subjective mind may be *en rapport* with any other subjective mind on this plane, or *on the spiritual*—because it is the soul.

A spiritualist medium being better developed than I, can come in touch with my subjective mind, when *both are passive,* and read for me what I cannot see or deduce for myself. But if I shut off my subjective mind by auto-suggestion, he might sit until Gabriel blows the trumphet and nothing will he see—though all the wild Indians and material mediums of the Irish world throng around him in droves. As if my departed loved ones needed an Indian brave to approach me! I asked a spiritualist once why they always used so illiterate a medium. She replied, "that Indians being so natural

in their lives clung closer to us and were floating nearest
earth, because not having progressed (?) on this sphere,
and having entered eternity, had not realized their
chance of progression over there." And yet, it seems to
me, they are especially favored since they can reach up
to any degree of heaven and bring back messages from
the more saintly. An Indian, "to raise a mortal to the
skies and draw an angel down!"

I set down nothing in malice nor will I use names,
for these mediums are well known around this city—
and Washington is very much given up to seeking the
occult through spiritualism.

There is one medium on whom I base most of my calculations. I will call him Mr. A., and I want to say, he
is an honest, true gentleman; a firm believer in his faith,
and altogether true to his convictions. I am very much
obliged for all the information gained through him.

In June, 1897, I made my first trial, merely through
curiosity, and not as an experiment. From this sitting
I gained but little personal information as I was much
interested in a friend's romance, and my mind seemed
filled with her desires, so my reading was all for her,
but I was promised better luck for myself the next year,
and a death by pistol shot, and some letters that I carried were interpreted, even the names of the writers.
These letters were held at the base of the cerebellum.
According to J. H. Dewey, this is the brain of the
inner sight. One card bore the compliments of an aged

friend, and this friend was promised me as a husband, and his parents in the spirit world stood near and besought me to accept him. When I add that this friend was bordering on seventy, lived in a little country village, and had lately buried his second wife, you can understand what a flattering prospect this was. My stars failed to ascend the next year; instead, they fought in their course against me, and yet who knows? It is always darkest before dawn, and I waded through blackest midnight before I reached the sunlight of the present. The pistol shot was authentic, and carried off a few months later a dear friend and companion—my beautiful collie dog.

In November of the same year I was passing through the city and much depressed in spirit over a seeming failure of life's forces; on the impulse of a moment, if we dare call spirit leadings impulses, I again visited Mr. A. He accurately and allegorically described the tumult I was under, and advised me to remain at home as a death was in my path. He described the little cemetery and even the name of the town. It is useless to say I acted on my own responsibility and did not stay at home. Six months later I was summoned to that home to see my dear father laid to rest in that little cemetery. Would you call this suggestion? *No.* He was reading my subjective mind, and I certainly never connected this prophesied death with my father, a man of robust physique and good health.

This death changed life for me. I took my aged mother, enfeebled in mind and body, entirely under my care—and in early June we moved to Washington.

Then I made my third visit to Mr. A., but with this difference, I had begun to develop my own psychic powers, and received my own impressions allegorically and clairvoyantly.

He described my mental conditions, told me of the influence of friends, and conditions I most wanted to know, and took up my life in the *same allegory* I myself used. Does this look like mind reading or spirit voices?

For the third time he interpreted letters for me, and failed to sense they had every time been from the same parties. Why? Because I shut off this information.

He told me of my father's death and of my mother's condition, and said, "When the leaves begin to fall, she will join him."

I leave the result of this sitting until later, and take up chance sittings of other mediums.

The next was a public hall and the medium a stranger. In these public seances articles are placed on the table to attract the spirits. Bosh! I used my mind and got my message.

She described the manner of my father's death, which, objectively, none of us knew since he passed away suddenly and in health, and had no organic trouble. Subjectively, of course, I knew.

She brought me this message from him; "That I

was vexed over a missive I had sent out and had re-ceived no reply, but it was all right, the silence meant good, and I would hear that week." Her reading was correct in every particular.

The next one was also public and the most amusing—twenty-five cents admission—private parlor, and three mediums present. We all sat around the room and waited. Some left when told a collection would be raised. The leader began to sing, but she was not fa-miliar with the words and could not find her glasses, so the song ended abruptly. Then a little man, with a great big book, began to promenade the floor, and fi-nally, with an humble apology for his nervousness, jerked out some words that were meant for a poem. They failed to reach my ears. Then there was a lengthy ad-dress by the leader, without any reference to grammat-ical construction, and it was all about the Spanish war and blood and all things horrible. I thought the spirits had taken flight, and wished sincerely I had, but it came to an end in time, and communications began in regular order.. Not one was passed ; each got his little message in regular rotation, but I don't know what mine was or from whence it came. I never heard of any of the spirits she brought me, but she explained I had so many around me she could hardly discern their names. She said I was planning three trips and not to go on any of them, which I most certainly did not, and she left me floating on a sea of doubt. Up rushed another medium, with finger pointed most menancingly, and told me I was worrying over an absent friend, but I would hear from him soon, or else see him. All very comforting but not realistic. Then came the little man, and being quite weary, I concluded I would try my power. He asked to hold my hand. I consented. He said, "You are not a spiritualist but you know something of the science;" then he shook his head and said, "I can get nothing, and you are the first with whom I have failed." He looked so heartbroken and ready to cry I almost relented, but drew him back to me the second time with same result. Then I did relent, and the third time I let him read my line of work. He seemed so pleased at his final success I told him the failure was not due to his lack of power, but I had shut him off. I was relieved when that meeting ended.

In October came a renowned, world-famed spiritualist to our city, and crowds flocked to his public seances. I received a very clear communication from him regarding my work, and I decided to test his private sittings— three dollars a test—and mine lasted ten minutes. When I entered the room, he said, "I see a great big 'S' over you." "Yes," I said, "perhaps you do; I have just been discussing the merits of the S. A. L. Railroad." "Oh, yes," he replied, "that is correct, for thoughts are things."

I don't recall anything he told me, excepting that death was not near me for twelve months, anyhow. (Mark this!)

Afterwards, I visited a great titled medium who had studied in India with the Mahatmas. I know but little of Theosophy, but I thought these worthy rulers of thought did not mingle with the world or directly receive students—perhaps my ignorance may be excused—but he did say he had traveled with Herrman, and did do many tricks of jugglery. Poor fellow! I am sorry for him, as I heard lately he had injured his eyesight, and his power had departed. Well, I wrote ten questions, enclosed them in an envelope, and kept them in my hand. He answered nine and gave them to me in writing, and all are wrong. I was to marry a very dear friend who already has a wife. A death was to occur on February 10th that would make changes for me. I am glad to say this is April and that death has not taken place. He promised my mother would be left me until 1900.

The first week in November I visited a celebrated lady medium. She talked very nicely about broken anchors and precipices and will power and God. She sent me north and she sent me west and prophesied a change in my condition about November 25th.

About two weeks later, my mother, sitting by the window, said, "All the leaves are off this tree." I looked up startled at the coincidence (?) of her remark. By the next noonday she was gone.

Mark the result of the June reading of Mr. A. Was this suggestion? *No;* else all the others would have

sensed this condition. With the departure of this dear
spirit a change in my own forces came, not by my own
conscious act or will, but from that great subjective
mind that rules all life when allowed to develop. Remember, for six months I had watched over and guarded
and cared for this life that was gone. She had lived on
my mentality day and night with no intermission and I
had gone to the verge of eternity with her, and when the
mortal body was laid away I still carried the spirit and
knew it not. When all was over I returned to the city
and was thoroughly prostrated. I lay for days almost
lifeless, not grieving, for death is a happy release to the
soul anxious to go, and it is a selfish sin to wish back
those who have won the victory. My mind was clear
but the body was lifeless. I had given all my magnetism those hours of waiting, and was utterly depleted and
crushed by a heavy load, but I was ignorant of the
cause. My mother seemed nearer to me than in life. I
would turn often and often to render her some little service. In the night I would get up and go to her bed
and suddenly remember after some struggle she was not
there.

One night a voice spoke and said, " Consult a clair-
voyant, a Mr. N., a stranger in the city." The next
day I went to him. He said, " Why have you come to
me? You need no spiritualist to read for you because
you get your own interpretations. You have a power
of your own and I can't get you an individual message

for you are too closely surrounded by thoughts belonging to others." Then I asked of the mother who had passed over two weeks before. Listen! He said, "It is too soon; she has not yet realized herself; she is living on your mentality, and you are carrying her spirit on the other side just as you carried it here, but you must have a wonderful power that you have so far aided her progression for me to even sense this condition in so short a period. Now use your power still further and send her spirit to its recognized plane."

For a week after I never ceased my suggestions, and on Sunday afternoon the conditions broke and a blissful, divine harmony flooded my entire body, and I knew she was free and with the other dear loved one.

Six months previous, when the father died, I was away from home and only reached there forty-eight hours after life had gone out. When I stood by his body I became subjective, and felt his spirit touch mine, and I said, "Dear father, in the spirit land, lead and guide me into light." I felt then an answering glow, but knowing nothing of spirit forces, and thinking nothing of life beyond the body, I did not follow up this tie, but I remained in that old home three weeks as a mortal, physical coward. My nerves were completely unstrung, and I was positively afraid to be alone, even in the day time. I had brought down on my head a force of which I knew nothing. Being a thoroughly material man, of great intellect but no spirituality, he lingered

there in his old haunts for a long time. I felt no fear
of ghosts and never imagined I saw anything, but I
was conscious of the supernatural all around me. I
wish I had known then what I know now. My father
and I had always been closely associated in all the business ways of life and we most closely touched in intellect. Now, when the mother joined him I fought
against using their forces. Life had not been all harmony, and I was glad for them to be free from the
trammels of the body and I did not want to link them
with the earth thought, but it was to be.

My life seemed tangled and mismated. I so often
saw them near me in my dreams and my soul would cry
out in its loneliness and yearning. For two months I
fought against acknowledging spirit forces, but I knew
I had not grasped the keynote of Truth. Just as surely
as we are creatures of circumstance and knocked around
by the mental atmosphere of association and material
thought, just so surely is this spirit of ours—this great
subjective mind—in communion with other subjective
minds that are freed from the slavery of the body, and
the two forces will war against each other until harmony
is established. The natural law in the spiritual world
must be acknowledged. One day, in greater despair
than ever, I laid down my arms and said, "I give up.
the fight. If you can reach me, lead me; I am weary
of the earthly struggle and I will follow this to the end."
Then my soul went up into the mountain top, and I was

transfigured by the joyful radiance of the light thrown round me. Did I stay there? No, for my work was in the valley, but from that day on I have been at peace and I can ascend into that holy place at will and gain counsel along the roughest places.

I will not say that life is an unruffled calm. I do not wish it so for that would mean a stand still, and my lesson is not all learned and will not be while I am in the body. I do not take the stand that most scientists do and claim to be invincible. While in the body I am subject to its external laws, but Truth can make me rise superior to its bodily ills. I will not say I do not, know aches and pains. I profit by them and I conquer them.

Not long since I asked an M. D. for a prescription for a local complaint. He said, "Don't use drugs; you don't need them, and they will do you no good." Nor do they; so when I faint by the wayside I retire into the silence and call my forces to my aid. The power that flows through me for the good of others is existent still for me. I must naturally be depleted at times. The body gives out according to its laws.

Jesus wearied and used no earthly means for restoration, but sought the silence of concentration and rebuilded himself.

In the beginning I was treated by suggestive therapeutics and lifted from a state of suffering and invalidism. Now, having individualized my forces, suggestion

from another, to all appearances, fails. I have demon-
strated this in myself by shutting off every symptom of
pneumonia. It was no easy matter and required many
days, but in that time I used no drugs and sought no
physician. I was determined to conquer or die in the
attempt. I conquered, and knew when material forces
broke. It is not right nor helpful to others to claim
we know not pain, because Truth has come to us. I
am telling no secrets of our science by citing a few in-
stances. One of the strongest men I know—a learned,
skilful scientist, succumbed to la grippe. A patriarch
of one of the most spiritual schools, that aims at immor-
tality in the body, was injured by a street car. He
lingered, suffering greatly—at last called in a celebrated
mental scientist, for he was too depleted to recover lost
power, although several months had elapsed since the
accident. She effected an almost immediate relief, produced sleep, and was most hopeful. Two weeks later I
learned he had passed away and was cremated. Was
he depleted of his spiritual power? No, but the physical body was. Seventy years had been given him and
the spirit was ready for its further progress.

A great healer in the extreme south and a great healer
in the far north were friends. A child of the southern
woman fell ill. None of that school could reach it, but
the northern mind succeeded.

"What was wrong?" I asked. The reply was, "what
is mine will come to me; what is yours I cannot touch."

Perhaps this theory is all correct. I do not say. I
will not take the stand another writer does and say, " I
am the way." I am daily progressing and am open to
conviction, and other ways may be better than mine.
But this I know; the gift of healing is not of me, but
some spirit force acting on and through my comprehension. And what is the center of all spirit force? God.

Occasionally some bright and shining light that has
gained prominence among men steps to the front and
with a clarion call speaks for freedom, and the public
press takes up his words and echoes them far and wide.
This is one great wave of vibrating force and lifts
humanity many degrees higher.

The Rev. Mr. Chadman, of New York, has lately elec-
trified his church and stepped beyond the limits of the
Methodist creed by boldly announcing the fallibility of
the Bible. Bold, courageous man ! Dr. Lyman Abbott
has come to the front for spiritism, and says, "I love to
think my mother follows me with her eyes as she did
when I was a boy. I love to believe that the strange,
subtle, inexplicable, and indefinable influence that
sometimes comes into my life is from her."

From the hour I accepted this control, life has been
brighter and happier. I am never alone, and I find my
wishes and suggestions interpreted with a strange power.

There is much mystery attached to the phenomena,
and there are days when conditions are better than
others and I get clearer results.

Always my father materializes first, because he is stronger and more akin to the earth, and our brain power was similar. The mother was more spiritual, but the two combined complete the sphere and are one.

My father often appears in the way I knew him best on earth and shows me leadings by written words. The mother appears in visions or allegories, and often the voice is distinctly heard, though in a slow, monotonous, far-off intonation. A blessed privilege is mine! And I dare not be sceptical and doubt the phenomena.

Lately, I attended a public seance of my former medium, Mr. A. It seemed farcical. I had been worried that day over personal matters, and sitting there, I said, "If I am not guessing at shadows send me an answer through this medium." I tried to draw him to me. Three times during the evening he started in my direction and each time was strangely held back. I got no message, but that night my father came to me in my sleep and answered me directly. Several days after I learned the result was correct.

Boldly had I stood forth and said, my spirit friends need not come to me through illiterate mediums, when they wanted to communicate with me they could do so directly. Was this suggestion? Of the very strongest. Now, mark this. I said in the beginning that I got my best results from Mr. A. Since accepting spiritism I have made him four visits, and he has not been able to even try and read for me. At the last visit I said, well,

I will come no more; I am satisfied; and because he is so honest I told him why he failed. He frankly admitted the wonder.

There is so much charlatanry, so much trickery and fraud thrown around the science that the world must necessarily look with doubtful and questioning eyes. I will not use my power and forces for fortune telling or arranging sentimental scenes. My loved ones are too dear for such vulgar work. I read for no one. But the gift of healing for bettering the condition of others is mine to give—and the stones I carve out for myself are given readily and gladly that others may climb higher on the steps I make.

There is one curious phenomena I must put in motion. The spirit forces and communications come quietly, softly, and sweetly. I awaken in the middle of the night and know they are with me—no fear, no nervousness.

But I hold a telepathic communication with one in the body, and though many hundred miles intervene, space is nothing. Visions and impressions come as clairvoyance and clairaudience, but at times there is a most curious magnetic manifestation. I am always awake, and there are generally three vibrations, each wave growing stronger. It seems to be a hissing, burning flame of electricity, and my body rocks and vibrates on this electric wave. I am able to carry on a laughing commentary of the whole proceedings. On the last occasion I caught within my hand the warm, living flesh and blood hand of my friend, and felt for a ring I knew the owner wore—not there—then I took the other hand and found the ring and twisted it around. Now, the strangest part of this phenomena is that the other party is not conscious of any act, but arises in the morning with simply an impression of something having occurred. I get up, bristling with electricity, have much trouble arranging my hair, and can send a shock through the body of another person even though he be a sceptic. On this last manifestion I said nothing, but wrote to the friend asking about the ring. The reply came, "I am having trouble with my finger and cannot wear the ring. In washing my hands soap blistered the flesh under the ring and I cannot get it over the knuckle of the finger of the other hand."

I wrote back, "You see the effect, not the cause."

I venture no explanation of this phenomena. I have had four other similar communication always with this same friend. Each time the manifestations are more distinct. Both of us possess strong magnetism. A sensitive, developed, cannot readily enter my circle without an electric shock. In experimenting, subjects say the impression is of a volt entering the subjective mind and passing to the objective knowledge along the sympathetic nerve system and produces a tingling sensation over the entire body. I make the experiment simply by a telepathic suggestion, and I find by development I have only to give the suggestion and apparently forget it.

In copying this manuscript for the publisher I use heavy paper. My other chapters were easily copied, but in writing of the spirit forces, I have brought their conditions around me, and the magnetic current is so strong that I have much trouble separating the sheets as I write.

There are laws within laws and we may none of us seek to fully unfold the mysteries. The struggle is not yet over but the vibrations are widening each day, and the time will come when we will truly say, "It is well with my soul."

IV.

BUILDING-STONES.

I have gathered a posy of other men's flowers and nothing but the string that binds them is my own.—
Montaigne.

We are all sculptors and painters, and our material is our own flesh and blood and bones. Any nobleness begins at once to refine a man's features; any meanness or sensuality to imbrute them.—*Thoreau.*

Divine wisdom in man does not speculate or "draw logical conclusions," neither is it dependent for knowledge on communications received from anybody, but it is the power of the true, living *faith;* i. e., the power of the spirit of man to grasp spiritual truths existing within its own self.—*Franz Hartmann, M. D.*

The activity of the universal mind can only come to the consciousness of those whose spheres of mind are capable of receiving its impressions. Those who make room for such impressions will receive them. Such impressions are passing in and out of the sphere of the individual mind, and they may cause visions and dreams having an important meaning and whose interpretation is an art that is known to the wise.—*Paracelsus.*

Serene, I fold my hands and wait,
  Nor care for wind or tide or sea;
I rave no more 'gainst time or fate
  For lo! my own shall come to me.

I stay my haste, I make delays,
  For what avails this eager pace?
I stand amid the eternal ways,
  And what is mine shall know my face—
                              John Burroughs.

Such as are thy habitual thoughts, such also will be the character of thy mind; for the soul is dyed by the thoughts.—*Marcus Aurelius Antoninus.*

All reform aims in some one particular to let the great soul have its way through us; in other words, to engage us to obey. There is no bar or wall in the soul where man, the effect, ceases, and God, the cause, begins. The walls are taken away. The soul circumscribeth all things.—*Emerson.*

As in the Christ, so constantly in us the lower life has to meet all dangers and all agonies—the hunger, the thirst, the weariness, aye, even the scourging and the cross—when the purposes of the higher call for it.— *Phillips Brooks.*

High hearts are never long without hearing some new call, some distant clarion of God, even in their dreams, and soon they are observed to break up the camp of ease and start on some fresh march of faithful service. And looking higher still, we find those who never wait until their moral work accumulates, and who reward resolution with no rest; with whom, therefore, the alternation is instantaneous and constant; who do the good only to see the better, and see the better only to achieve it; who are too meek for transport, too faithful for remorse, too earnest for repose; whose worship is action, and whose action ceaseless aspiration.—*J. Martineau.*

We are like to Him with whom there is no past or future, with whom a day is as a thousand years, and a thousand years as one day, when we do our work in the great present, leaving both past and future to Him to whom they are ever present, and fearing nothing, because He is in our past, as much as, and far more than, we can feel Him to be in our present. Partakers thus of the divine nature, resting in that perfect all-in-all, in whom our nature is eternal too, we walk without fear, full of hope and courage, and strength to do His will, waiting for the endless good which He is always giving as fast as He can get us able to take it in.—*G. MacDonald.*

Do not think it wasted time to submit yourself to any influence which may bring upon you any noble feeling.— *J. Ruskin.*

Our souls crave a perfect good ; we feel the pull thitherward, we own the law that points in that direction.—*Anon.*

The voice of the soul is not to be silenced.—*Anon.*

The thinking being thinks its own conditions into the world outside of it.—*Anon.*

Looking down hinders, even though the intent is to escape evil.—*Anon.*

People grow like what they look at; that is eminently true of those who look at the highest.—*Anon.*

Build new domes of thought in your mind and presently you will find that instead of your finding the eternal life, the eternal life will find you.—*Anon.*

Thou hast made us for Thyself, O Lord, and our heart is restless until it rests in Thee.—*St. Augustine.*

If thou wouldst have aught of good, have it from thyself.—*Epictetus.*

Without hurry and without rest, the human soul goes forth from the beginning to embody every faculty, every thought, every emotion which belongs to it, in appropriate events.—*Emerson.*

There is a time in every man's education. when he arrives at the conviction that envy is ignorance; that imitation is suicide; that he must take himself for better, for worse, as his portion; that though the wide universe is full of good, no kernel of nourishing corn can come to him but through his toil bestowed on that plot of ground which is given him to till.—*Emerson.*

Trust thyself. Accept the place the divine providence has found for you, the society of your contemporaries, the connection of events.—*Ibid.*

No law can be sacred to me but that of my own nature.—*Ibid.*

What I must do is all that concerns me, not what the

people think. This rule, equally arduous in actual and in intellectual life, may serve for the whole distinction between greatness and meanness. It is the harder because you will always find those who think they know what is your duty better than you know it. It is easy in the world to live after the world's opinion; it is easy in solitude to live after our own; but the great man is he who in the midst of the crowd keeps with perfect sweetness the independence of solitude.—*Ibid.*

The power men possess to annoy me I give them by a weak curiosity. No man can come near me but through my act.—*Ibid.*

Another sort of false prayers are our regrets. Discontent is the want of self reliance, it is infirmity of will. Regret calamities if you can thereby help the sufferer; if not, attend your own work and already the evil begins to be repaired.—*Ibid.*

Insist on yourself; never imitate. Your own gift you can present every moment with the cumulative force of a whole life's cultivation; but of the adopted talent of another you have only an extemporaneous half possession. That which each can do best, none but his Maker can teach him.—*Ibid.*

Nothing can bring you peace but yourself. Nothing can bring you peace but the triumphs of principles.—*Ibid.*

Every man in his lifetime needs to thank his faults.— *Ibid.*

There is a deeper fact in the soul than compensation,

to wit, its own nature. The soul is not a compensation, but a life. The soul *is*. Under all this running sea of circumstances, whose waters ebb and flow with perfect balance, lies the aboriginal abyss of real being. Essence, or God, is not a relation or a part, but the whole.—*Ibid.*

What your heart thinks great, is great. The soul's emphasis is always right. Take the place and attitude which belong to you, and all men acquiesce.—*Ibid.*

The life of man is a self-evolving circle, which, from a ring, imperceptibly small, rushes on all sides outwards to new and larger circles, and that without end. The extent to which this generation of circles, wheel without wheel, will go, depends on the force or truth of the individual soul.—*Ibid.*

Nothing great was ever achieved without enthusiasm. The way of life is wonderful; it is by abandonment—*Ib.*

Life is a lesson. Count all joy, all pain no more than part of what the soul must learn in this great school, the world.—*Grace Macomber.*

> Blindfolded and alone I stand
> With unknown threshholds on each hand:
> The darkness deepens as I grope,
> Afraid to fear, afraid to hope.
> Yet this one thing I learn to know
> Each day more surely as I go,
> That doors are opened, ways are made,
> Burdens are lifted, or are laid
> By some great law unseen and still
> Unfathomed purpose to fulfil.
>   "Not as I will."—*H. H. Jackson.*

If you'll sing a song as you go along,
In the face of the real or the fancied wrong;
In spite of the doubt, if you'll fight it out,
And show a heart that is brave and stout;
If you'll laugh at the jeers and refuse the tears,
You'll force the ever-reluctant cheers
That the world denies when a coward cries
To give to the man who bravely tries;
And you'll win success with a little song
If you'll sing the song as you go along. —*Anon.*

Luck is the tuning of our inmost thought
To chord with God's great plan.   That done, ah, know
Thy silent wishes to results shall grow,
And day by day shall miracles be wrought.
Once let thy being selflessly be brought
To chime with universal good, and lo !
What music from the spheres shall through thee flow !
What benefits shall come to thee unsought !
Shut out the noise of traffic !   Rise above
The body's clamor !   With the soul's fine ear
Attune thyself to harmonies divine.
All, all are written in the key of Love.
Leap to the score, and thou hast nought to fear.
Achievements yet undreamed of shall be thine.—

Ella Wheeler Wilcox.

V.

MORTAL MIND.

" Yet all experience is an arch where through
   Gleams that untraveled world whose margin fades
   Forever and forever as I move.
   How dull it is to pause, to make an end,
   To rust unburnished, not to shine in use;
   As though to breathe were life.   Life piled on life
   Were all too little, and of one to me
   Little remains; but every hour is saved
   From that eternal silence.   Something more—
   A bringer of new things; and vile it were
   For some three suns to store and hoard myself,
   And this gray spirit yearning in desire
   To follow knowledge like a sinking star
   Beyond the utmost bound of human thought."

This manuscript was folded, ready for the publisher,
when a voice said to me, " wait;" and I had to wait.
Circumstances compelled me.   No matter which way I
turned, I was nonplused.   Finally, a great wave of experience swept over my psychic brain showing why I
was stopped, and I dared not go on without giving the
intuitions and guidings to the public, since all are seeking
beyond the bounds of human thought to follow knowledge—and yet the untraveled world extends like its
sinking star.   Poets live in the psychic realm.   They
touch the emotions of life and give voice to the silent

thoughts of every nature. Perhaps I quote too freely, but the language expresses the vibrations I want to send forth, and it may be only one snowy feather of truth falls, but it is worth the trouble of trying even to send forth that atom.

Whoever thinks to lay down a working hypothesis that all may follow, and follow as a given rule good for any length of time, is mistaken. Nature changes constantly. Progression never stands still, and a psychic nature must move through all experiences and not rust unburnished. You cannot possibly tell yourself you have found a truth and expect to hold it. You will outgrow it and rise to other heights. No, I admit this is no blissful state, nor a restful one; but it is true living. From the cradle to the grave is one complete and increasing progression. What we think this week we may not think next week, nor to-morrow. Certainly not, if we be thinking people. Everything changes—customs, people, governments, religion. As well expect to put a bar across the earth's orbit as expect to concentrate your mind on one set rule.

The planets and their satellites are in constant motion, and it has been proved that the sun itself is not fixed, but all obey one grand law. And yet man lays down a rule and says, " Here, follow this, or be outside regulation thought."

The world is running mad after its desire for intangible substances, and it will not be satisfied ; neither will it be satisfied after it passes the chasm death makes until it rests in its own knowledge of progression.

With all their seances, all their tests, all their materializations, spirit controls give us no idea of the life beyond.

They tell us they are happy and at rest. But do they give any idea of heaven as taught by the churches? Of the personal God they have worshipped, or of Jesus, their mediator? Has anyone ever returned and said he was like Dives? God forbid!

Loving messages we get, guidance along our earthly path—questions answered, but it is all so unsatisfactory, and we spend our time chasing shadows, and are no nearer the truth.

Psychic phenomena are varied and interesting, but they are lurid, gleaming darts that carry us over rugged paths. That there is something within the reach of all, the world feels, but what that something is, the filmy veil of material thought shuts away from mortal vision. Every individual considers it a compliment to be told he or she is magnetic or may be a medium. What does it mean? Everyone does possess magnetism to some extent or he would not be here. Everyone is a medium by his attracting force, only some use it differently from others.

The hypothesis of the duality of mind is seemingly correct, but there comes a time when questions arise. After death seizes the body what becomes of the objective mind? Does it become submerged into the subjective, although they are distinct and separate entities?

Who has passed through the dissecting room and held within his hands the brain of a human being? Truly wonderful it is with all its complications sending out its nerves and nerve force, and its still more wonderful convolutions. Yet who of us dare say this mass of gray and white matter thought and acted of its own volition? If so, why so useless after death? What caused the convolutions? I clipped the following from a daily paper:

"Dr. Hanseman, of the Berlin University, has made an examination of the brain of the late Prof. Helmholtz. The average weight of a man's brain is 1358 grammes. Gauss' brain weighed 1492, Cuvier's, 1600, and Helmholtz's only 1440, but in its frontal part it presented a very unusual development and number of convolutions, and it is these convolutions, not brain weight, that modern physiology associates with intelligence."

Now, what is intelligence? We rightly say animals have intelligence and yet we call it instinct?

In human beings, reason holds sway, and instinct is classed as intuition, or over-wrought imagination.

A vain multiplicity of words that satisfy not, and still we go on seeking for light.

What does it matter about the future if we know how to live in the present? And to live is not only to breathe, but to extract all the sunshine, all the joy, all the health, that is meant to be our inheritance.

I sat by my window the other night through many hours watching engineers at work on an electric road, and thought how little they knew of the force they were then controlling.

Nor do we know the force we think we control. Mortal mind!

Some one has boldly said there is no objective mind, it is all subjective. We know mind is a force; what else need we know? Mind is not the brain, but it is the power that controls the nerve force, that produces the convolutions that endow us with life, that lives after the body is dead. If there be two minds, what becomes of the other? One law pervades all nature.

All the rivers and streams in some way reach the ocean. All the oceans are connected, forming one vast body of water with its numerous and intricate network of outlets.

One mind, one force, one law, guides and controls the universe. All human beings are evolved manifestations of materiality pervaded and controlled by the one mind according to our hereditary ideas of suggestion.

We are individualities reflecting the one Divine mind, and we do not know ourselves. Do not know our birthright, our privileges—but live as slaves to our surroundings and are too phlegmatic, too indolent, to seek out our own salvation and carve our own channels.

Individuals we are not, but units of a mass, while content to plod along the line of least resistance.

One evening last week I saw by the paper there would be a double hanging the next day. I thought no more of it. At least, I didn't think I remembered, forgetting, if such be possible for long, my subjective mind.

The next day was cloudy and heavy. A great oppression rested on me. I could not rid myself of it. It was not mine, but what was it? All day it grew in intensity. Something seemed trying to impress itself on my conscious thought. Some truth seemed hovering. I was curiously lead even to an undertaker's shop to telephone to a neighboring city on business, and the paraphernalia of death made my nerves keenly sensitive.

But still I could not unravel the mystery. Later in the afternoon, I went, strangely enough, to another portion of the house to talk with a patient. Her first words were, "I shall be so glad when this day is over and this hanging is a thing of the past."

The mystery was revealed. Every one I met that day bore the same testimony—psychics and materialists —the dead oppression of the mental atmosphere.

"All are but parts of one stupendous whole." The sudden ending of those lives caused a reaction—a jar of the entire universe—and those nearest felt the strong first vibrations! An injury to one effects the whole.

Where was, and what is Mortal Mind?

We do not comprehend its powers any more than we can explain electricity.

A phenomena has lately come to my experience that I feel justified in giving to investigating minds. Led by curiosity and some invisible force I visited a slate-writing medium—a man altogether different from the usual class of mediums.

I asked if he believed in his science?

He honestly replied, "I certainly do, but I cannot explain it or even understand it."

Now, the results would have been the same had I carried my own slates.

We sat on either side of an ordinary table over which was thrown a chenille cover. I wrote three questions to different persons in the spirit world and folded each paper separately. The medium did not see the questions.

I washed off those ordinary school slates with a wet sponge, dried them with a cotton cloth, tied two slates together with my own handkerchief, and held them several inches above the table. The medium also had his hands on the outer edge. He had broken off a fine point of a slate pencil and placed between the slates.

The spirits did not immediately respond. He insisted I write one more question to balance the sexes. So I added to the list a very material friend who had passed away about two years ago.

I will give my questions in detail and the replies. The slates I brought away with me and have in my possession.

1. C. P. McCabe: How far wrong am I in my occult work? Is there anything you wish to communicate to me?

2. Margaret A. McCabe: Are you happier than on this sphere? Are you at rest?

3. Fannie Robins: How will my affair end?

4. (By the medium's request) John Robins: Shall I go to Norfolk on Sunday?

Suddenly there came a scratching sound between the slates as of a pencil writing. Now mark the reverse order of communication and the signatures.

1. Little by little I am learning what real life is. I do not feel that I ever lived until I got into this sphere. The spirit is the real being, not the body, and in or out of the form it lives on and is the same. I never expected to see you here. Yes, go to Norfolk on Sunday if things be shaped for it properly there.

J. L. ROBINS.

(This penmanship is identical with the writer's style and the signature he always used. I did not go to Norfolk, however, and don't see that I lost anything.)

2. Good morning. How charming it is to come here
and find that one can be received and welcomed by you.
It is so hard to bear the rebuffs of those we love in
such meetings as these. I know indeed full well how
difficult it is to realize the verity of these things when
one has come up along religious lines, so averse to the
acceptance of a revelation so wonderful as this. I
thank you for coming here and making my soul happier.
Your affair will end to your way of wishing.

Lovingly,

FANNIE ROBINS.

MY DEAR CHILD: Receive this little letter as a
token of my love and regard. I am not gone from you.
I still live and see and know all. Think of me as I
am, alive and well and comparatively happy. I am
happy and at rest.

Devotedly, Mother,

MARGARET A. McCABE.

(These two on one slate and arranged in this unique
way. Signatures perfect, but penmanship strange and
both written in a different hand.)

3. This is indeed a suprise and a pleasure to meet you
here. How did you know I could be here and meet
you in such a strange way? Well, think no more of
me as one dead and gone. I am just as much myself as
I ever was. Your line of work and thought is correct.
Some experiences need modifying. I am at rest.

Father,

CHAS. P. McCABE.

Your mother is with me.

(This handwriting is not familiar but that may be explained by natural methods because in life he employed a secretary. The question arises—if the mother and father were together why did they not write on one slate?)

This phenomena is particularly interesting from its relation to mortal mind. If spirit hands penned those lines, they materialized sufficiently to employ ordinary means, but why after this writing, was the infinitesimal point of pencil not consumed? It was just the same size after filling three slates as when placed there, and this we know would be an impossibility if used.

These communications upset all other theories of the months previous.

These are not satisfying. Were the subjective personal communications, as related in my essay on Spirit Forces correct, why did these same spirits in writing me take this meeting as a surprise? And in coming so close to earth why did they confine themselves to simply answering little questions when weightier ones existed? And why was the father the last one to respond when, according to all spiritualistic lore, he should have been nearest, since it was the anniversary of his death and our combined thought brought him nearer the earth? There was so much sameness about the communications, so stereotyped, and the answers to my questions were most politic, throwing the results on my own forces.

So the query goes on and mortal mind performs its tricks and makes dupes of us beyond our ken.

On the night of the anniversary of father's death I retired to bed and to sleep. I was suddenly aroused from the heavy sleep to the sub-conscious state by a heavy knock against my shoulder. I said, "Who is it? Is it you, father?" and tried to turn over to see, conscious of some presence back of me but I could not turn. I was held down by strong pressure. The struggle to arouse myself and shake off the oppression resulted in over excited nerves and I gained consciousness with an effort. In common parlance this may be described as an ordinary "nightmare," only I must confess I am not subject to these hallucinations, this being my first experience, nor had I eaten anything previous to retiring since my usual dinner hour, six hours before, and my digestive organs are perfect. But that night was remarkable, for every article of furniture in my room kept up its incessant raps. Something unusual; besides the room was filled with magnetism. Determined to investigate, the next night I made preparations to receive any spiritual visitors that might come. Removed all portable furniture into another room, especially willow rockers that would naturally rap, according to philosophical vibrations, even if the electric vibrations of thought did not reach them, gave myself strong treatment against nerves, laid down and went to sleep, and slept soundly until morning. Now, how can

I account for the previous night ? Easily enough. Mortal mind acting on nerves. In the room directly under mine lives a couple who devoutly believe in spiritualism. They are not professionals and are limited in their investigations. The woman is constantly experiencing raps and thumps and, having lately conversed with her, my mortal mind must go to work and produce all this phantasmagoria for my delectation. Now the vibrations of my restlessness were felt by my materialistic (?) friend at a distance of several hundred miles. This morning I received a letter saying, "What was wrong with you Saturday night? I was with you in thought but could get only an uneasy impression, besides I could not sleep and in the morning I said, 'Now, what was she up to last night?'" Mortal mind again, for never once did my thoughts wander to that distant friend consciously.

It may have been spiritualistic rappings, but I want nothing so material. I claim the majesty and dignity and Godliness of the subjective mind, and if in the spirit world the souls of my departed loved ones wish to guide me they can and will use natural means and not use contracting fibers of wood to let me know they are near.

If the concentrated belief of spiritualists have established this code of signals satisfactory to themselves it is not for me to deny their experiments and pleasant belief, but if my psychic brain, being *en rapport* with their mental atmosphere, produces similar results, why the next night was my belief and power of auto-suggestion so strong as to make me perfectly oblivious of any demonstrations even were spirits near me? Still I am not wedded to my ways and am open to conviction, only I cannot take table raps and furniture blows to mean spirit messages when I have a mind in connection with the Universal Mind.

So many people gain words and think they have the science. I have had so many questions put to me on the power of auto-suggestion that I feel called to explain here what I mean.

Repetition of words, denying impressions, exercising the will, are not tokens of auto-suggestion. To do this one must be able to concentrate thought and pass into the sub-conscious state, which is easily recognized once you have gained the knowledge and then give yourself the suggestion.

This is followed by a feeling of exhilaration, lightness, buoyancy, freedom of the body and magnetism or electricity.

This is an adjunct of every individual, but it is not every one who takes the trouble to develop because it requires patience, perseverance, courage, and fearlessness.

Ordinarily it can't be done; for there are necessary conditions to be observed and a control of the entire nervous system. It is not developed momentarily, though many achieve greater results than others who labor longer and more untiringly, but the power once obtained is the keynote of health and universal good and after development it may be practised almost unconsciously—by a mere wish or desire—while the conscious brain is performing regular duties or conversing along other lines.

There are pre-natal impressions that strengthen this development, or oppose it. There are special gifts and there are evils blacker than night that follow in its wake if the control be not pure.

Of stage work and hypnotic passes, to please and amuse curious people, I know nothing and will not exercise them. Their day is over and the harm they have thrown around this God-like science is degrading.

I hypnotize by telepathy, and the magnetic current thus emanating soothes the nerves and brings refreshing sleep. To use such gifts for any but healing purposes is devilish.

Suggestion is a power passing the knowledge of man and suggestion is mind in its magnetic currents operating on kindred minds less developed or equally so and expectant.

I think the entire trouble lies in our selfishness, which is due to our material conception of life. We want our own sphere to be one of happiness and contentment, peace and luxury, and do not give of our sympathy and love to all individuals.

We lack tenderness and affection except where it pleases us to bestow it.

If I could call back six months of my life, when I did not know how to show tenderness, did not know how to use my powers and give what yearning human hearts wanted, I would gladly live over those months, and they cover the only period of my life that I would voluntarily go through again. I was so blindly living, only for the best life, suited to fill my idea of enjoyment. Now that the difference has come and I realize I am only a reflection of all life, I want to link myself with individual miserable existence and bless as I pass. My heart seems bleeding when I watch the misery around me.

One of the saddest sights met my view to-day. A good, honest, little colored woman, seized with the craving for drink, left her work and went out on the street. A few hours after I saw her lying on the floor, completely under the influence of liquor, with a bunch of dead roses beside her that she had picked up. She was one of God's creatures and a part of the universe.

And this deadly curse can be entirely removed by the power of suggestion.

Church people and scoffers are lifting their hands in holy horror at what they call the blasphemy of mental scientists and yet they say they believe their Bible, when all its teachings show the one Universal Mind, as God, and spirit.

Jesus said, "I and the Father are one." How?

And again, "Unto the least of these as unto Me."
What do they mean by their spiritual teachings?

There is no mystery, but there is love, and healing,
and happiness, and health.

Mortal mind is an adjunct of the body until the
tissues of the body run their day; then mortal mind becomes its spiritual self. It returns to its origin, and
being a portion of the Universal Whole, it can produce
all the phenomena of its own imagination, and by its
own desires and credulity can make us slaves to material life, or give us freedom in the joy of illumination.

Lead kindly light, amid the encircling gloom,
    Lead thou me on.
The night is dark and I am far from home,
    Lead thou me on.
Keep thou my feet, I do do not ask to see
The distant scene; one step's enough for me.

I was not ever thus, nor prayed that thou
    Shoulds't lead me on.
I loved to choose and see my path; but now
    Lead thou me on.
I loved the garish day; and spite of fears,
Pride ruled my will. Remember not past years.

So long Thy power hast blest me, sure it still
    Will lead me on.
O'er moor and fen, o'er crag and torrent till
    The night is gone.
And with the morn those angel faces smile
Which I have loved long since and lost awhile.

VI.

ALPHA AND OMEGA.

The beginning and the end. One divine completeness. Eternity! Starting from any point of the circumference, with unvarying result, we arrive always just where we begun, but we had the joy of progression and of knowing the circle, which we never should have done had we not set out on the race, and the one great truth would have been unknown to us. There never was any beginning, there never will be any ending.

There is only God. There never was and there never will be anything else. I am God. You are God. We came from Him, and when this mortal body has run its day we return to God. Nothing is lost. Everything is just as it was when the Word spoke, and creation began. God never retracts, never takes back, and what was spoken remains, intangible, invisible, useless probably, but not by God's intention, only because of man's perverted, ignorant eyes. When passion is cooled and the purblind race to the grave is ended, physical eyes will be opened to the vast unfathonable truth that surrounds us, just within the grasp of every living being, but separated by greater walls than ever China raised around her kingdom; walls of structure so enduring, towering up beyond and hiding the sunlight of God's love, shutting out the secret of true living, making each individual life of such limited proportion that bounded by his own small circle he cannot feel the soul-communing of the life around him, and this wall is simply ignorance and narrow reasoning belief.

Better the "crackling of thorns beneath the pot" than the pessimistic Pharisaical wall of intellect which refuses to see and be true to its own convictions.

The physical senses are only sentinels of mortal mind. We do not carry them with us beyond the grave. The soul needs no avenue for its functions.

God is everywhere.

> " From Nature's chain whatever link you strike,
> Tenth, or ten-thousandth, breaks the chain alike."

You cannot strike any link from Alpha to Omega because you can't find a link. It is just one perfect harmonious whole. The mistakes and troubles that arise are products of mortal mind. God never meant any such complications, but man, wise in his own conceit, has built up his own wall and shut God out; and man is rudderless and yet does not comprehend where the trouble lies. Let but one assault be made on that wall, the slightest channel be cut, and the soul allowed to spring into consciousness, and I defy any craft to ever try and entirely obliterate that still, small voice from whispering to its own and gathering in force until by mighty friction the victory is won, and that man stands forth as a representative god, uniting and blending the perfect attributes of material, intellectual, and spiritual attainments.

Of all characters that ever swept across this world's stage, none have exceeded in brilliancy and daring the hero of Austerlitz. Born of obscure parentage, he united ambition and will, and by the greatest strokes of diplomacy wielded a power under which Europe trembled. His early youth is described as taciturn, wilful, studious, and a dreamer. By development he rose to the height of his prominence; by invincible courage and steady determination he yielded not to any whisperings of defeat, but backed by his arrogance and vanity, though crushed to the earth by the enemy, he rushed forward to his final overthrow; and, sadder than the sufferings of Josephine, deeper than the agony of the brave army on its retreat from Moscow, the field of Waterloo gave forth the cry of the conqueror, crushed to rise no more. Self-love defeated! The dreamer lost in his own vain imaginings because the connecting link of the wildest, strangest, strongest mentality was cast aside by mortal pride!

Hugo prefaces his description of the great charge by these pathetic words: "Had it not rained on the 17th of June, 1815, the future of Europe would have been changed. A few drops of water, more or less, prostrated Napoleon. That Waterloo should be the end of Austerlitz, Providence needed only a little rain; and an unseasonable cloud crossing the sky sufficed for the overthrow of a world."

Just a cloud sweeping across the horizon may make or mar history! Is this coincidence or intention? Happy the man who understands the laws of life and can utilize that cloud for the upbuilding of his own forces. The defeat of one is the victory of another, but the happy medium of equalizing magnetism produces a calm that floods all eternity with sunshine. Nature is soothing and speaks always with a voice like music to a soul sick with selfishness.

Tired and weary with the cares of a life that seemed aimless, weary of the heart-yearnings and passionate longing for a life beyond the commonplace existence of mere living, a mortal traveling through the hill countries of a northern State had a message from nature, interpreted by this same still, small voice. All around was ice and snow; cold, barren hillsides, and valleys slumbering at their feet, until awakened by their Sun-God; bleakness and desolateness, but on the hilltops the reflection of the sunshine.

In this mortal heart were colder, more icy blasts. Every desire, every ambition, every wish seemed frozen and covered by a pall deeper than the snowdrifts and perhaps not so pure, and if the sunlight of God's love had broken its way through the walls of self-structure it was unfelt.

Down from the mountain top a little stream had made its path, winding in and out among the rocks and gathering force as it swept over the boulders, until as it crossed the road it had formed one broad track—but silent now—frozen into one solid beautiful example of patience and waiting. Under that mantle of ice life waited and listened for the voice of its Sun-God to bid it sparkle and flow, singing on its course. The lesson sank deeply into that frozen, embittered, lonely heart, and the soul pulsating beneath that icy barrier sprung into life and found its answer to the problem of living.

Aye, friction is necessary. We find it in the human framework. We find it in all great constructions, but there is an antidote. There must be friction always, until the soul gains its kingdom, and the oil upon troubled waters has robbed life of its tempestuous waves.

Years ago, in a woman's college, one of the students carelessly dropped a chemical she was holding. The liquid fell on the front of her dress directly over the heart. The upright beam fell three inches and the cross beam measured two inches. The woman was a psychist and fearless. For four years that cross was an omen.

"Misfortunes came not in single spies but in battalions."

Directly before any great affliction or sorrow that emblem came to the subjective mind as a vision, and everywhere did the word "cross" appear. It was the

Irish Banshee! I am not commenting on the law of "suggestion" now, but simply stating the subtle working of a curiously wrought psychic brain.

On Palm Sunday she was surrounded by crosses made of reeds, every scholar wore one, and many were presented to her. Before the week ended afflictions multiplied, in less than a month one parent passed suddenly into eternity and within six months the other followed, and every earthly tie of the old life was swept away. Everywhere she looked the emblem of the Crucified One cast its shadow upon her. Mortal self was warring heavily with the ego of her inner consciousness. Friction most unaccountable pressed heavily, but the soul grew on. One night she saw a vision of a little wayside chapel, and through dreamland she sought an answer to the absence of the accustomed 'cross upon its tower—but no answer came, and the cross was conspicuous by its absence. She had learned to await development of allegories, which are only pictures intuition paints upon the soul's canvas. Next morning, driving down the city by a church she had passed before without consciously observing, there suddenly stood out in its golden splendor against a cloudy sky—the cross : and the dream was answered.

What had come as Alpha in a vision was lost in the Omega of another vision.

The cross was only mortal suffering, trials, and afflictions deeper than words can portray—but when the

Christ came into its own, bright and shining was the cross against its dark background of woe and clouds.

Think you I am dreaming dreams, and seeing visions? Nay, I speak truth.

The world is slowly awakening to a great joy, unspeakable and full of glory, a joy that has always been —but not understood. Invisible cords connect all life, for there is but one life. Vibrations reach out from the center to the periphery. Nothing is lost, though many are too blind to see. Time and space do not exist, only to mortal mind, and the life that pulsates and vibrates within me links and joins me with all life, no matter what material space intervenes. The great subjective mind of all eternity separates nothing from itself.

Telepathy is an interesting experiment to curious seekers after phenomena, but to those who know its laws it is simpler than telegraphy. Nearly all are trying to find its secrets, but it has none.

There is one Universal Intelligence and from its source all life draws its power.

By concentration and desire states of consciousness become one. The question is, not whether these states communicate with each other, but whether mortal mind has the power to draw this result above the threshold of consciousness. Some have gained this power by persistent and patient effort, and possibly by friction. Patience and perseverance work out all results. The silence of Vibrating Intelligence makes plain all mortal crosses, makes them shine forth with splendor greater than the noonday sun; robs life of its ills, its heart-aches, and its misunderstandings.

Learn the only secret nature holds—the secret of patience—eternal repose !—and all things are yours.

" Clouds, that in their very motion breathe of rest," the "murmur and glistening" of life; the "instinct reaching and growing into the soul of a flower"—breathe nothing but repose.

Silence—the silence of God—the Holy of Holies—where mortal mind comes into harmony with the Divine Mind, and the ego asserts itself!

Learn this secret power, and learn then the controll-ing force of soul communing. Learn the Alpha and Omega of life. Learn to recognize God and yourself.

As He was in the beginning, is now, and ever shall be; as He was before the Past began, as He will be when the Future is ended.

www.ingramcontent.com/pod-product-compliance
Lightning Source LLC
Chambersburg PA
CBHW021423090426
42742CB00009B/1225